TURTLES
AROUND THE POND

written and photographed
by
Mia Coulton

Here is a turtle.

3

Here is a pond.

A turtle is sitting on a log in the pond.

Turtles like to sit on logs on sunny days.

5

Here is a turtle walking on the road.

7

Here is a turtle walking in the grass.

Here is a turtle hiding in the grass.

11

Here is a turtle hiding in its shell.